I0158147

F-84 Thunderjet

Republic Thunder

Hugh Harkins

Copyright © 2013 Hugh Harkins

All rights reserved.

ISBN: 1-903630-61-4
ISBN-13: 978-1-903630-61-7

F-84 Thunderjet
Republic Thunder

© Hugh Harkins 2013

Published by Centurion Publishing
Glasgow
United Kingdom
G65 9YE

ISBN 10: 1-903630-61-4
ISBN 13: 978-1-903630-61-7

This volume first published in 2013

The Author is identified as the copyright holder of this work under sections 77
and 78 of the Copyright Designs and Patents Act 1988

Cover design © Centurion Publishing & Createspace

Page layout, concept and design © Centurion Publishing

All rights reserved. No part of this publication may be reproduced, stored in a
retrieval system, transmitted in any form, or by any means, electronic,
mechanical or photocopied, recorded or otherwise, without the written
permission of the Publishers

The Publishers and Author would like to thank all organizations and services
for their assistance and contributions in the preparation of this publication

CONTENTS

INTRODUCTION   v

Chapter 1   XP-84 & YP-84A   6

Chapter 2   F-84B/C/D/E/G   16

Chapter 3   KOREAN WAR   43

APPENDICES   52

GLOSSARY   56

INTRODUCTION

While no US designed jet aircraft saw operational service during World War II, many would be employed in the 1950 – 1953 Korean War, including the Republic F-84 and the earlier Lockheed F-80, both of which were employed in large numbers. The F-84 was designed during the latter stages of World War II, known then as the XP-84 (The USAF changed from the 'P' for Pursuit to the 'F' for Fighter designation in 1948). The aircraft was redesigned and did not fly until 26 February 1946, with the first production variant, the P-84B (later F-84B) attaining an Initial Operational Capability in December 1947.

Designed primarily as a fighter, the introduction of more advanced fighters like the North American F-86 Sabre and the fielding by the Communist Air Forces of the swept wing MiG-15 saw the F-84 quickly outclassed in the air to air arena during the Korean War, which began on 25 June 1950, lasting until 27 July 1953.

F-84's were deployed to Korea in November 1950, primarily as escort fighters protecting Boeing B-29 Superfortress Bombers. However, high bomber losses to MiG's led to the USAF abandoning the daylight bombing role for the B-29, after which, F-84's increasingly moved more and more to the air to surface role.

CHAPTER 1

XP-84 & YP-84A

The Republic XP-84, above, adopted a relatively simple design approach with a simple nose air intake feeding the turbojet engine. AFFTC

The Republic Corporation became a major name in the US fighter aircraft industry with the beefy P-47 Thunderbolt, which along with the North American P-51 Mustang, was the mainstay of the USAAF (United States Army Air Force) fighter forces in the later stages of World War II. With the world of aviation taking strides into the jet era during World War II, the USAAF wanted to introduce a number of tactical jet fighters to operational service to reduce the gap with Britain, which fielded the jet engine Gloster Meteor from July 1944.

Development of the Republic XP-84 began in 1944, and three XP-84 prototypes were ordered in March 1945, with the design being refined in the months following the end of WW II in August that year. The XP-84 emerged as a simple straight wing aircraft powered by a single General Electric J35-GE-7 (license built British de Havilland Goblin) turbojet engine rated at 3,750-lb (7.6-kN) static thrust. Unlike the Lockheed P-80 Shooting Star & Bell P-69 Airacomet, which had lateral fuselage mounted intakes, the XP-84 adopted a simple nose mounted intake, which fed air directly to the engine housed in the rear fuselage with the jet pipe at the extreme rear in a similar fashion to the earliest German and British jet aircraft, the Heinkel 178 and Gloster E.28/39.

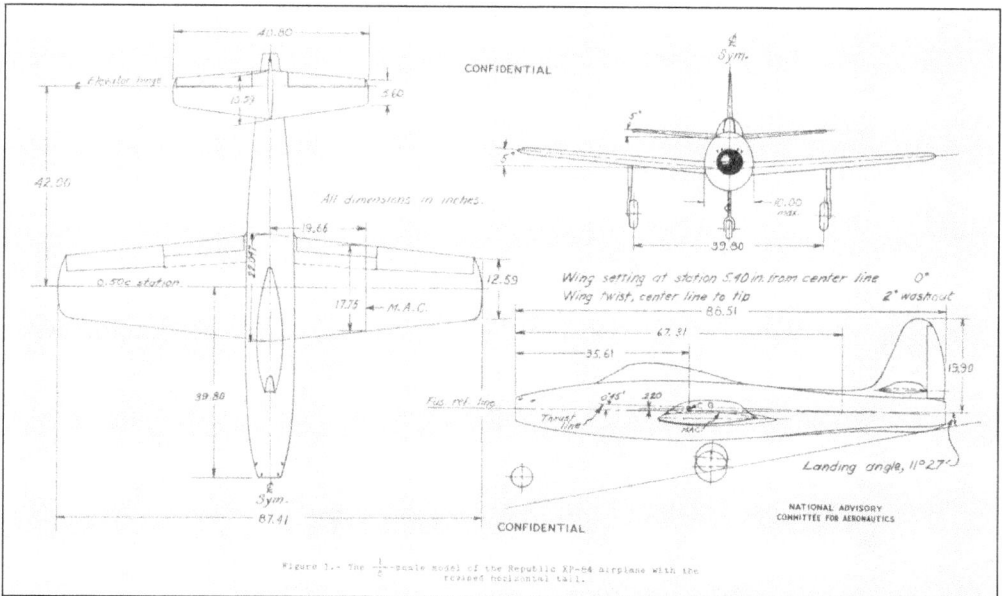

Top: NACA conducted preliminary tests of a Buffet Stall-Warning Device on this 1/5 Scale Model of the Republic XP-84; here mounted in the NACA Langley 300 mph 7 x 10 feet wind-tunnel, with turbulence net in position. NACA

Above: Three view of a 1/5 model of the XP-84 fitted with a revised horizontal tail. NACA

XP-84 S/N: 45-59475, which was the first of the XP-84's built, clearly shows the simple straightforward design. The design of the aircraft commenced in 1944 and the XP-84's were ordered in March 1945. These aircraft were powered by the General Electric J35-GE-7 turbojet engine rated at 3,750-lb (7.6 kN) static thrust, which was basically a license built variant of the British de Havilland Goblin turbojet. USAF

Top and centre: XP-84 S/N: 45-59475. Bottom: Head-on view of an XP-84 showing the nose air intake for the J35 engine. USAF

9

Top: The first XP-84, S/N: 45-59475, during flight testing. This aircraft conducted its maiden flight on 26 February 1946. During its early flight series it was discovered that there was a "complete lack of stall warning", fixes for which were developed during tunnel test at NACA Langley.

Above: The second XP-84, S/N: 45-59476, conducted its maiden flight in August 1946. USAF

The first XP-84, SN: 45-59475, in flight shows the aircrafts bulky forward fuselage and the almost 'X' like view, courtesy of the un-swept wings, when seen from above or below. The aircraft was powered by a J35-GE-7 turbojet engine rated at 3,750 lb static thrust. This unit was a developed licence built variant of the British de Havilland Goblin turbojet used to power the de Havilland Vampire twin-boom single engine jet fighter then entering service with the RAF (Royal Air Force). USAF

GOR (General Operational Requirement), issued on 11 September 1944, specified a day fighter featuring mid-set wings, with a top speed of 600 mph, a combat radius of 850 miles armed with either 8 x 0.50 calibre or 6 x 0.60 caliber machine guns; this requirement later being reduced to either 6 x 0.50 or 4 x 0.60 caliber guns. The aircrafts required radius of action was also reduced to 705 miles. The go-ahead was ordered on 11 November 1944, followed on 4 January 1945 by a letter of contract for a USAAF order for 25 service test aircraft and 75 production aircraft designated P-84. The order was later changed to 15 YP-84A's, and 85 P-84B's. On 12 March 1945, a definitive contract was signed for three aircraft designated XP-84, along with other developmental items including a static test model. The contract was modified in June 1945 to incorporate the previous letter contract, which was now dropped.

The aircraft suffered many development problems leading to a modified design being put forward in July 1945, with the third XP-84 to be built as the XP-84A.

Top: The second XP-84, SN: 45-59476, during a test flight. This aircraft, which flew for the first time in August 1946, was identical to the first XP-84, S/N: 45-59475. AFFTC

Above: The 2 XP-84's and the single XP-84A were followed by a batch of YP-84A aircraft procured for service test and evaluation. YP-84A SN: 45-59483 was typical of the 15 YP-84A's procured. USAF

Top: YP-84A S/N: 45-59517. Above: YP-84A S/N: 45-59495. Deliveries of YP-84A's to the USAF commenced in February 1947. These aircraft were powered by the more powerful Allison J35-A-15 turbojet rated at 4,000-lb (8-kN), and unlike the XP-84's, were armed with a battery of 6 x 0.50 in M2 machine guns. USAF

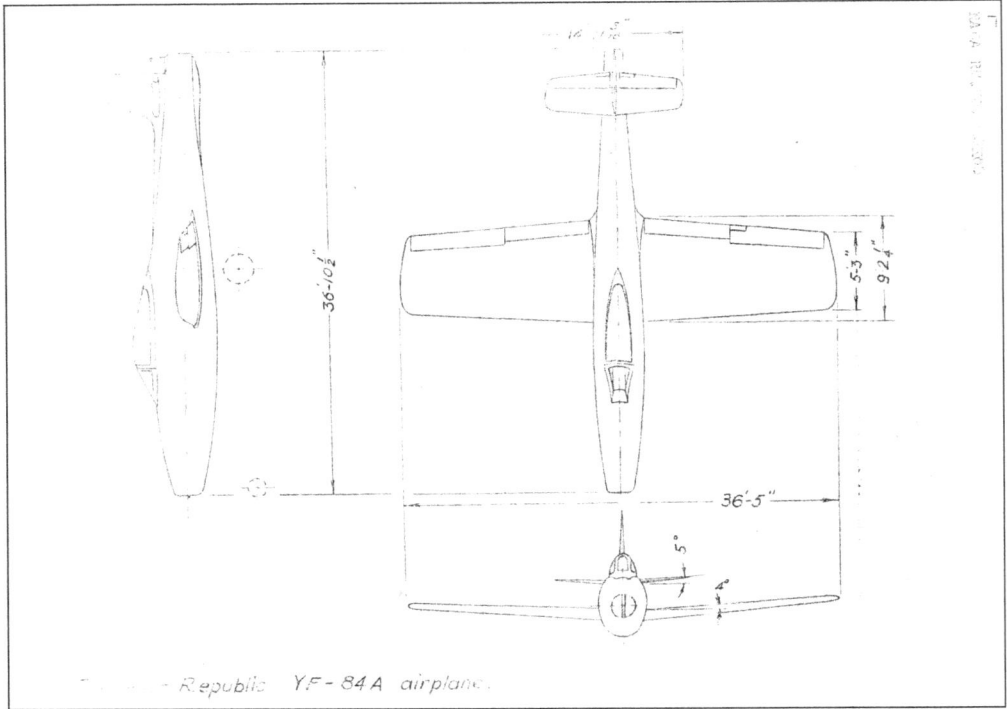

Three-view general arrangement drawing of the YP-84A (top) based on aircraft S/N: 45-58488 (above), which was used for test Measurements in Flight of the Longitudinal-Stability Characteristics of the YP-84A. NACA

Above: YF-84A S/N: 45-59488 during service evaluation with the USAF. USAF

The first of the XP-84's ordered took to the air for the first time on 26 February 1946, with second flying in August that year, both powered by the J35-GE-7 turbojet rated at 3,750 lb static thrust, and this XP-84 J35 combination was used to set a new US national speed record of 611-mph (983-km/h) in September 1946. Despite the speed record flight, the design suffered from weight problems and was underpowered, a problem which would effect a number of production variants.

The two XP-84s were followed by the single XP-84A and a batch of 15 YP-84A's (initially 25 YP-84's were planned), which were powered by the Allison J35-A-15 turbojet rated at 4,000-lb (8-kN); these being used for USAAF service evaluation, with the XP-84A being completed more or less to YP-84A standard.

The YP-84A's were delivered to the USAAF in February 1947 for service evaluation. Unlike the XP-84's, these aircraft were armed with 6 x 0.50 in M2 machine guns; four in the upper forward fuselage area and two mounted in the wings. The YP-84A's were also equipped to carry wingtip drop tanks; the additional weight being compensated for by the more powerful engines.

The NACA Langley Aeronautical Laboratory operated Y-84A S/N: 45-59490 and the Ames Aeronautical Laboratory operated YP-84A S/N 45-59488. Both these aircraft were eventually transferred to the NACA HSFS (High Speed Flight Station) co-located at Muroc (later Edwards AFB) in November 1949 and December 1950 respectively. 45-59490 was used for some research flights and as a chase aircraft and as a pilot proficiency hack, while 45-59488 was used as a spares source for the first aircraft.

CHAPTER 2

F-84B/C/D/E/G

The P-84B (later F-84B) was the first production model of the Thunderjet. The PS- code was used when aircraft were designated 'P' for Pursuit, but was changed to 'F' for Fighter in 1948. USAF

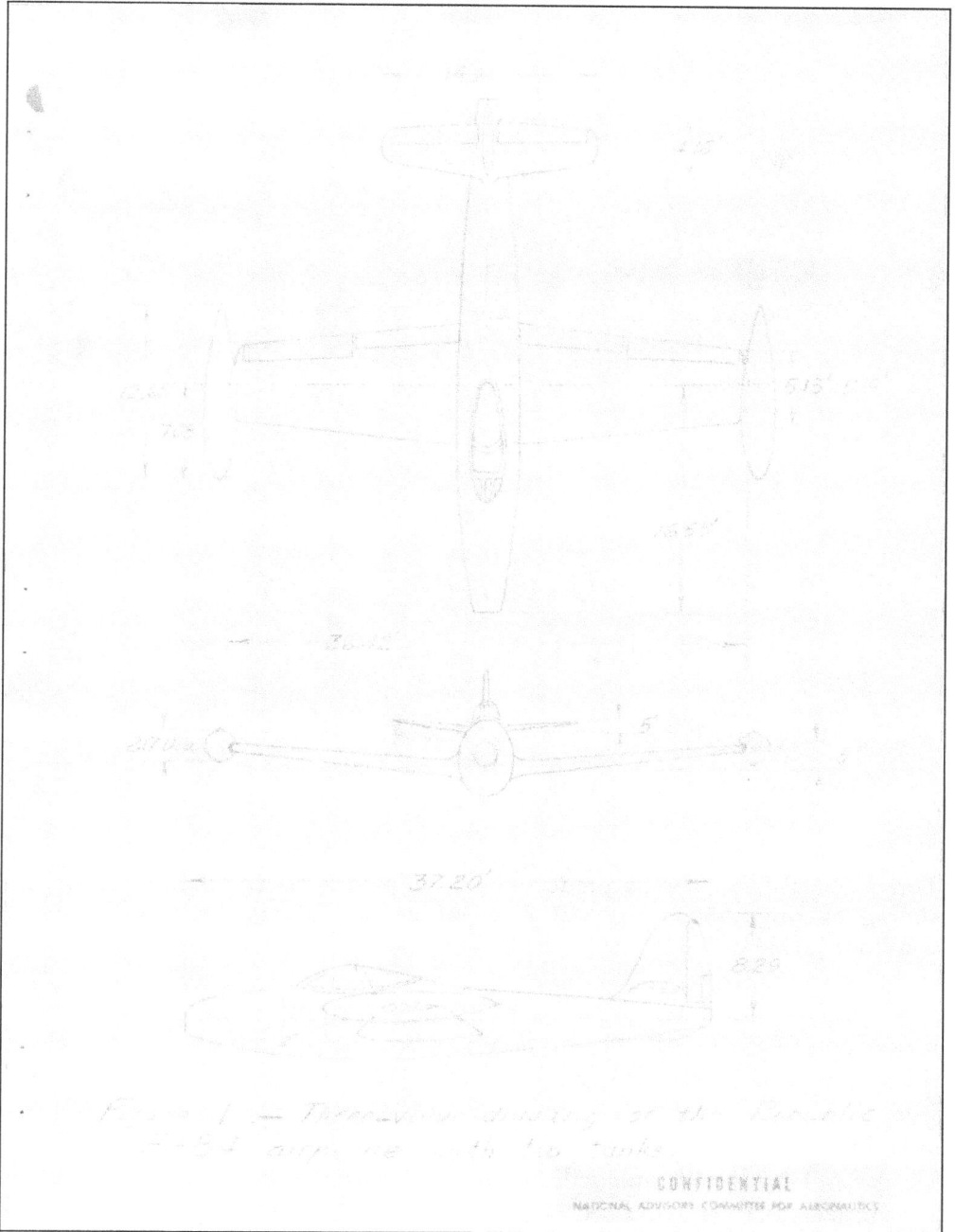

This very faint three-view general arrangement drawing of the Republic P-84 (F-84), showing the basic dimensions of the aircraft, comes from a 1940's test document. NASA Ames Research Centre

Top: The P-84B, like its predecessors, adopted a relatively straight forward sliding canopy. Above: P-84B 45-59561 at an air base in winter weather conditions. USAF

A contract for the production of 99 P-84A's was cancelled in favor of production of the P-84B, which, therefore, became the first production variant, the first of which was completed in June 1947. The P-84B, the first of which were delivered to the USAAF in summer 1947, was more or less an YP-84A armed with the M3 variant of the 0.50 in machine guns instead of the M2 in the YP-84A. The P-84B attained an IOC (Initial Operational Capability) in December 1947 with the 14th Fighter Group based at Dow Field, Bangor, Maine. This IOC was, however, subject to a number of serious flying restrictions due to ongoing problems with the aircraft. The ongoing structural problems with the P-84B saw the entire fleet grounded in May 1948 for inspection, after which aircraft were returned to flying status, albeit with many

18

restrictions, while modifications were designed. In 1949, a program of structural modifications was introduced, including strengthening the wings, and around 100 or so other structural enhancements. From the 86th P-84B, delivered in late 1947, the M3 machine gun armament was supplemented by eight rockets mounted in retractable launchers on the underside of the wings.

The cockpit of a P-84B was typical of fighter cockpits for the period. USAF

The lack of fuselage space and thin wings meant that the P-84 could not carry much in the way of fuel internally; therefore, wingtip fuel tanks were fitted to production aircraft. The P-84B was equipped with an ejection seat. Other changes included introduction of the J35-A-15C engine, which was rated at the same 4,000-lb thrust as the -15 engine of the YP-84A.

The last P-84B was delivered in June 1948, by which time 226 aircraft had been accepted; less than half of the aircraft ordered under various contracts. The balance of these contracts would be passed over to later variants. On 11 June 1948 the USAF changed the P (Pursuit) designation to F (Fighter) resulting in the P-84 being re-designated F-84. The F-84B was phased out of the USAF inventory by the end of 1952.

Top: F-84B's undergoing maintenance. Above: A formation of F-84B's during a formation flight over Main, Eastern USA. USAF

Top and above: P-84C's. This variant introduced a number of minor improvements over the P-84B and was powered by the J35-A-13 engine rated at 4,000-lb (8-kN). USAF

Basically an improved F-84B, the next production variant was the F-84C, 191 of which were delivered, featuring some improvements over the F-84B, including a more reliable electrical system and the J35-A-13 engine rated at 4,000-lb (8-kN). The first 11 F-84C's went to the 20th FG (Fighter Group) in May 1948, with later deliveries going to the 33rd FG. The 31st and 78th FG's operated F-84C's along with F-84B's. The last F-84C was retired from active USAF service in 1952.

Top: F-84C S/N: 47-1589 on the ramp at an air base in the Continental United States. Above: F-84C-16-RE S/N: 47-1580 in service with the Illinois ANG in 1955. These aircraft are equipped with the heavily framed cockpit canopy. USAF

Top: A trio of F-84D's from the Connecticut ANG. Above: An F-84D from the 79th Fighter Bomber Squadron, 20th Fighter Bomber Wing on the ramp at Eglin Air Force Base, Florida in 1950. Carried over from F-84C contracts, the F-84D introduced a number of changes over the F-8C which preceded it, including redesigned wings and a more powerful J35-A-17D engine rated at 5,000-lb (10-kN) static thrust. USAF

Next page: An F-84D in zoom climb over the Californian desert, while serving with the AFFTC (Air Force Flight Test Centre) during the early 1950's. AFFTC

The F-84C was followed by the F-84D, 154 of which were built from F-84C contracts. This variant introduced more radical changes, including new wings with a thicker wing skin gauge, a longer fuselage, re-designed undercarriage and a more powerful J35-A-17D engine rated at 5,000-lb (10-kN) thrust (some records show that some F-84B's were powered by the J35-A-13 engine used in the F-84C), which was fed by a fuel system optimized for cold-weather operations.

The F-84E was the second most numerous variant, with 843 aircraft built, 100 of which went to Mutual Defense Assistance Program customers. This variant, the first of which conducted its maiden flight on 18 May 1949, introduced a number of changes, including a lengthened fuselage. USAF

Top: F-84E Cockpit. Above: F-84E S/N: 49-2028 serving with the Wright Field Test Division. USAF

This F-84E is carrying a heavy load (for the F-84) of 12 x 5 in HVAR (High Velocity Aircraft Rockets) on the wings and a pair of Tiny Tim air to surface rockets on the shoulder stations. The 'E' was also armed with the standard six gun battery of M3 0.50 in machine guns. This load out on 21 October 1949 was rather more than would be carried when the F-84's was used operationally just over a year later. USAF

F-84E, S/N: 49-2124, conducts a rocket assisted take-off courtesy of the RATOG (Rocket Assisted Take-Off Gear) pack located under the rear fuselage just aft of the wing trailing edge. RATOG could be used to significantly reduce the take-off run of an aircraft, particularly when heavily loaded. USAF

The 1947 contract that led to the delivery of 191 F-84C's was amended to cover delivery of an additional 154 aircraft designated F-84D, the first of which was delivered to the USAF in November 1948, with an additional 36 aircraft delivered before the end of that year, and with all 154 being delivered by the end of April 1949. The F-84D commenced its retirement from the active USAF inventory from late summer 1952, but served on with the ANG (Air National Guard), which operated the 'D' until 1957.

On 29 December 1948, the F-84 program was overhauled and a new contract provided for production of 409 F-84E's. The F-84E, powered by the J35-A-17 engine, introduced many changes, including a lengthened fuselage, stronger wing, more spacious cockpit, a radar gun sight and improvements to the fuel system, which included the wingtip fuel tanks, and provision for 230 US Gallon tanks under the wings, increasing the aircrafts combat radius to more than 1,000 miles.

Top: Eight F-8E's from the Ohio ANG. Centre: This F-84E is configured with wingtip and shoulder station mounted external tanks. Above: Two ANG F-84E's in the 1950's. USAF

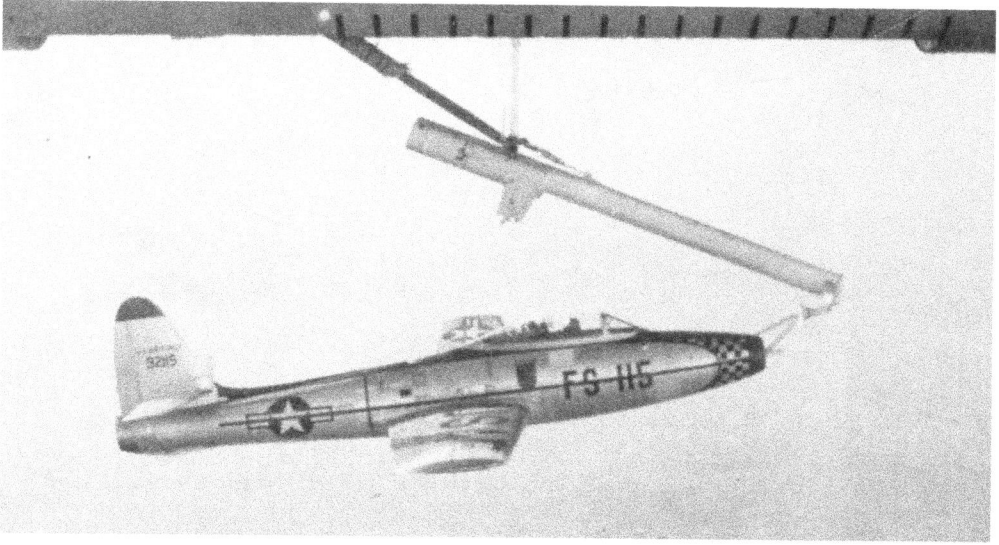

Previous page: The F-84E in the stowed position on the B-36. This page: The F-84E retrieving position– long boom configuration (top) and short boom configuration (bottom). USAF

The first production F-84E conducted its maiden flight on 18 May 1949, and was, along with another aircraft, accepted on the 26th of that month. Testing showed that the 'E' model was satisfactory in basic performance and maintainability, but problems arose with the A-1B sighting system leading to a suspension of deliveries until the A-1C sighting system became available in early 1950.

The F-84E in the extended position hung beneath the B-36 Mothership. USAF

Various contracts saw 743 F-84E's delivered to the USAF and a further 100 delivered to MDAP (Mutual Defense Assistance Program) customers; the last 3 aircraft being delivered in July 1951. The F-84E served with the USAF until 1956, at which time it was serving with TAC (Tactical Air Command) as a training aircraft. The Air Force Reserve relinquished its last F-84E's in 1957 and they were retired from ANG service in 1959.

FICON

An F-84E was used in the development of the FICON (Fighter Conveyor) program. The B-36/F-84E combination was known as the 'Bomb Bay' type of Parasite system. This involved a trapeze system "used to support, launch, and retrieve the parasite during flight… The sequence operations of the trapeze is controlled from the carrier; this control is vested in an operator located in a pressure capsule in the bomb bay… Operational equipment for two-stage type missions are provided."

The contract for the prototype FICON, RB-36F/F-84 combination (MX-1602), was awarded on 19 January 1951. The exploratory flight test program with the F-84E and the YF-84F, under which flight conditions under the bomb bay area of an RB-36D was evaluated, was completed on 31 March 1951 and the first RB-36/F-84E composite flight, with the F-84E housed in the bomb bay was conducted on 14 May 1954. Phase II flight testing, which was completed on 29 May 1952, involved 68 airborne launches and retrieve operations from the RB-36 carrier in 36.8 flight hours. On 15 October 1952, the RB-36F/F-84E Parasite combination was delivered to the Air Proving Ground for evaluation. Through 20 February 1953, there were 170 "arial launch and retrieve operations" in 280 flight hours.

It was never intended to produce an F-84E operational parasite variant. This variant was simply used to prove the feasibility of such a system for use with the RB-36/RF-84 swept wing parasite system for reconnaissance and bombing operations. 24 RF-84F's were eventually converted to RF-84F 'parasite' configuration, the system being used for a short period of time in the second half of the 1950's.

Top: This F-84G, like other tactical fighters of the period, is sporting a dazzling non tactical art scheme. Above: F-84G S/N: 51-9998 fitted with wingtip and shoulder mounted external tanks, which increased the aircraft's operational radius. USAF

Top: A quartet of Royal Norwegian Air Force F-84G's. The 'G' model was the most numerous production variant with **1,936** of the **3,025** production aircraft delivered to NATO nations under the MSP (Mutual Security Program). Above: The USAF received more F-84G's, **1089**, than any other previous variant (although figures from a number of documents vary.) USAF

The F-84G cockpit. USAF

Although designed as a jet fighter, by the time of the Korean War in 1950, the F-84 was more at home as a first line fighter bomber, a role in which it was heavily employed by the USAF during that conflict. Thunderjet's were also involved in air combat operations particularly as close escort for Boeing B-29 bombers, but were outclassed by CCAF (Chinese Communist Air Force) MiG-15 jet fighters.

Top: The F-84G was equipped for in-flight refueling, typically from USAF Boeing KB-29 Tanker conversions from B-29 Superfortress bombers using the hose and drogue system. USAF

Above: Once established in service the F-84G became the mainstay of USAF tactical fighter bomber Squadrons. USAF

Busy flight line scene at a Continental United States Air Base in the 1950's USAF

A major problem for the Thunderjet during its service career was its lack of engine thrust, resulting in the aircraft being underpowered, particularly when employed in the fighter-bomber role. The Thunderjet required a huge take-off roll even when carrying a modest bomb load, which saw the aircraft fly huge numbers of sorties, during the Korean War often with modest payloads.

To overcome the F-84's lack of performance, Republic embarked upon a swept wing variant, which emerged as the F-84F Thunderstreak. However, the urgent need for more power and a larger load carrying capability led to the development of an interim variant designated F-84G, which was basically an improved F-84D. This variant, was powered by the more powerful J35-A-29 turbojet engine rated at 5,600-lb thrust. While previous Thunderjet variants were most associated with conventional air to surface weapons such as rockets, bombs and napalm canisters, the F-84G was also tasked with theatre Atomic (later known as nuclear) strike carrying a single free fall atomic weapon, the first USAF tactical fighter to be employed in this role. Indeed, tactical nuclear strike was the primary role for this variant, although it could also be used as a conventional fighter-bomber. For operations with the free fall atomic bomb the F-84G was equipped with a LABS (Low Altitude Bombing System).

The ZELMAL was a rather crude attempt at overcoming the problem of lack of runway space. USAF

The F-84G introduced a new in-flight refueling system that allowed the aircraft to refuel from the new Boeing developed flying-boom refueling system being adopted by SAC (Strategic Air Command) , however the flying-boom compatible system did not appear on the F-84G until 1952. The 'G' featured an autopilot, A-4 gun sight (introduced from the 86th aircraft), a new Instrument Landing System (introduced from the 301st aircraft) and could carry 4,000 lb of stores, with the capability to deliver atomic weapons (this capability was introduced in late 1951).

The ZELMAL EF-84 during launch (top) and landing (bottom). The landing could be more accurately described as a controlled crash, with the arrester hook catching a wire, which then pulled the fighter down onto a large portable air filled mat. Ultimately the idea was flawed, for among other reasons in that it required specialist equipment, which would not have been available at forward bases. USAF

A typical load out for the Republic F-84 Thunderjet over Korea was a pair of 500 lb bombs. USAF

The first F-84G's were delivered in July 1951, and the aircraft entered service with SAC's 31st FEW (Fighter Escort Wing) based at Turner AFB, Georgia, which received its first aircraft in August that year. A total of 3,025 F-84G's were accepted, with 1,089 of these going the USAF and 1,936 going to MDAP recipients (some documents suggest the figures were 789 and 2,236 respectively).

The last production F-84G's, 21 MDAP aircraft, were delivered in July 1953. By August 1954, SAC had retired the F-84G's, but it remained in service with TAC; the last aircraft was retired from USAF operational service in mid-1960.

The F-84G was the first western tactical single-seat fighter to operate with tactical nuclear weapons, paving the way for future generations of supersonic nuclear armed strike fighters such as the F-100, F-101A/C and the F-105. The F-84G was equipped with an autopilot for long-range ferry flights, which were now part of operational doctrine, courtesy of in-flight-refueling. Operational restrictions caused by the flying-boom in-flight refueling method led to the F-84G being equipped with the hose and drogue system developed across the Atlantic by Flight Refueling in the United Kingdom. A pair of EF-84E's was converted in the UK and became the first jet powered fighter aircraft to cross the Atlantic non-stop courtesy of in-flight refueling on 22 September 1950.

In November 1950 the Fifth Air Force began receiving F-84's from the 27th FEG (Fighter Escort Group), the primary role of which was to protect Boeing B-29 Superfortress four engine bombers on daylight missions over North Korea. Outperformed by the more advanced MiG-15, the F-84 proved inadequate in the bomber escort role, as would the Gloster Meteor F Mk.8 twin jet fighters of the Royal Australian Air Force's No.77 Squadron a year later. Here F-84s from the 27th FEG are being lifted onto the aircraft carrier USS *Bataan* for the journey across the pacific to Japan. USAF

In the early 1950's, the vulnerability of NATO airfields to attack, which might result in the denial of runways to aircraft, was considered serious and solutions to the problem were sought. One solution was inspired by early cruise missiles such as the Regulus and Matador, which were launched from small ramps in a manner not unlike that used by Germany with the V-1 Flying Bomb during World War II. Having amassed experience on the Matador ramp launched cruise missile program, the Glenn L Martin Company was awarded a contract to develop a system, which could be used in conjunction with the F-84G fighter-bomber.

The F-84G ZELMAL (Zero Length Launch Mat) program involved launching an F-84 from a zero length platform on a rocket booster, which propelled the F-84 to flying speed. The program, which was surprisingly straightforward, worked well, with pilots on average suffering from no more stress during launch than that experienced from a catapult launch from an aircraft carrier. The landing process proved to be more of a problem as the

ZELMAL aircraft had its undercarriage removed to make launching easier. Instead of a conventional landing, the ZELMAL was equipped with an arrester hook, which would catch a wire and pull the fighter down onto a large portable air filled mat developed by Goodyear.

This F-84 is undergoing an engine change at Taegu airfield. USAF

Phase I flight-testing began with the launch of an un-piloted EF-84G from a trailer mounted launcher at a remote spot on Rodgers Dry Lake on 15 December 1953. The first flight of the piloted F-84G ZELMAL was conducted on 5 January 1954. During the first landing of the ZELMAL F-84G the aircraft caught the arrestor wire, which pulled the aircraft down onto its belly on the runway. However, the aircraft skidded off the runway. The second piloted launch was conducted on 28 January 1954, and, like the 5 January flight, was brought down to a landing on a conventional runway, with air mat landings not commencing until June that year. Although further landings worked better, support for the concept waned and the program was eventually cancelled.

The US ANG operated large numbers of Thunderjets, with some 14 ANG squadrons receiving F-84B's before the start of the Korean War in June 1950. When additional aircraft were required for Korean service a number of ANG squadrons lost their F-84's, although nine other ANG squadrons converted to the Thunderjet during the Korean War period. Post Korean War, Thunderjet's were operated by 10 ANG squadrons, with the last being retired in 1958. When retired from service some F-84 airframes were used for test purposes and as target drones with 80 retired F-84B's converted as target drones for the USN.

CHAPTER 3

KOREAN WAR

This F-84 adorned with 'Four Queens' art is armed with a 2 x 500 lb bombs.
USAF

F-84D/E's were deployed to Korea in December 1950, serving with the 27th FEW (Fighter Escort Wing). Due to substantial losses in fighter bombers, particularly during the railway interdiction campaign, combined with normal operational losses, Fifth Air Force began to receive additional F-84D's in spring 1952; 102 aircraft, most of which were allocated to the 136th Wing, which was a former ANG unit. Despite these reinforcements the F-84D was withdrawal from operational service as newer variants began to appear in numbers in August and September 1952.

The F-84E's inability to adequately protect the Boeing B-29 Bombers led to its gradual withdrawal from the bomber escort role; however, it performed adequately in the new role of ground attack and interdiction.

Eight B-29's from the 307th Bomb Group were tasked to attack the Namsi Airfield on 23 October 1951. As the bombers approached the target, MiG-15's attacked and all but swamped the F-84 escort and 3 B-29's were shot down. Most of the other B-29's in the formation were damaged and had crew members either killed or wounded. Although some documents give the date for this raid as 23 October, other operational documents point to the raid having taken place on 25 October as noted below, although the 23 October date is considered the more accurate.

Previous page top: The F-84 entered the Korean Theatre in December 1950. Here ground crews work on an F-84 in harsh winter conditions. Previous page centre: F-84's of the 27th FEG. Outclassed in the air to air role by the Russian MiG-15, the F-84 would eventually be used primarily in the interdiction and close air support roles. Previous page bottom: In the ground attack role the F-84 was a potent weapon armed with a variety of weapons. This aircraft is configured with 500 lb bombs and 5 in HVAR (High Velocity Aircraft Rockets). USAF

This page top: A formation of eight F-84E's during the Korean War, armed with 500 lb bombs. USAF Above: An F-84 takes off for a mission over Korea carrying a very modest bomb-load. AFFTC

Previous page top and centre: **F-84's are prepared for a ground attack mission armed with two 500 lb bombs. Previous page bottom: The air to air refueling probe in the front of the starboard wingtip fuel tank of this F-84E was utilized during Operation 'High Tide' to conduct long-range bombing missions against North Korean targets.**

This page: The F-84G eventually replaced earlier variants deployed to Korea. USAF

The following comes from the Monthly Tactical Reports emanating from Fifth Air Force Squadrons showing the inadequacy of the F-84 and the Gloster Meteor 8, when operating against the MiG-15 fielded by the Chinese Communist Air Force: **"On the 25th October 1951, a formation of nine B29s escorted by sixty-four F84 Thunderjets attempted to bomb Namsi Airfield. They were attacked by Migs and three B29s and 2 F84s were destroyed. Two B29s landed at Kimpo with major damage and the remaining four aircraft were damaged. Mig losses were light.**

On 29 October eight B29s, with sixteen Meteors on close escort, 64 Thunderjets on medium and high escort, and 32 Sabres as a screening force, attacked the railway bridge at Sinanju. The bombers were attacked after turning off the target and one was badly damaged. Only one flight of 12 got home an attack. No Migs were claimed by the escorting fighters."

Top and above: F-84 fuselage artwork during the Korean War. USAF

Following these failures to protect the bombers a conference was held at Itazuki, Japan on 28 October 1951. The results of this conference were that the bombers would fly at the lower altitudes of 12-15,000 ft, whereas they had previously attacked from 20-24,000 ft. Although this increased the threat from flak, it was considered acceptable as the straight wing fighters – F-84's and Royal Australian Air Force Meteor F.8's, were at less of a disadvantage against the MiG-15 at these lower altitudes. The F-86's would continue their role as a screen at a distance of some 30 miles "to one side of the bombers run-in at 25 to 30,000 feet".

Despite these changes the escorting fighters were still unable to protect the bombers, eventually resulting in Bomber Command moving to night time bombing, the F-84 and Meteor fighters then assuming an ever increasing air to surface commitment.

Over Korea F-84's were employed first as bomber escorts then primarily in the ground attack role. The Department of Defense reported 260 F-84's lost. F-84's from the 474th FBW en-route to a target north of the 38th Parallel in 1952. DoD

The F-84E, like previous models, despite being rated satisfactory in regards to maintenance, was plagued with operational problems, which at times were so acute that something in the order of 50% of the USAF F-84 inventory was non-serviceable. An example of this is the fact that of the 60 F-84E's deployed to the FEAF in December 1950, only 27 aircraft were actually fully operational capable.

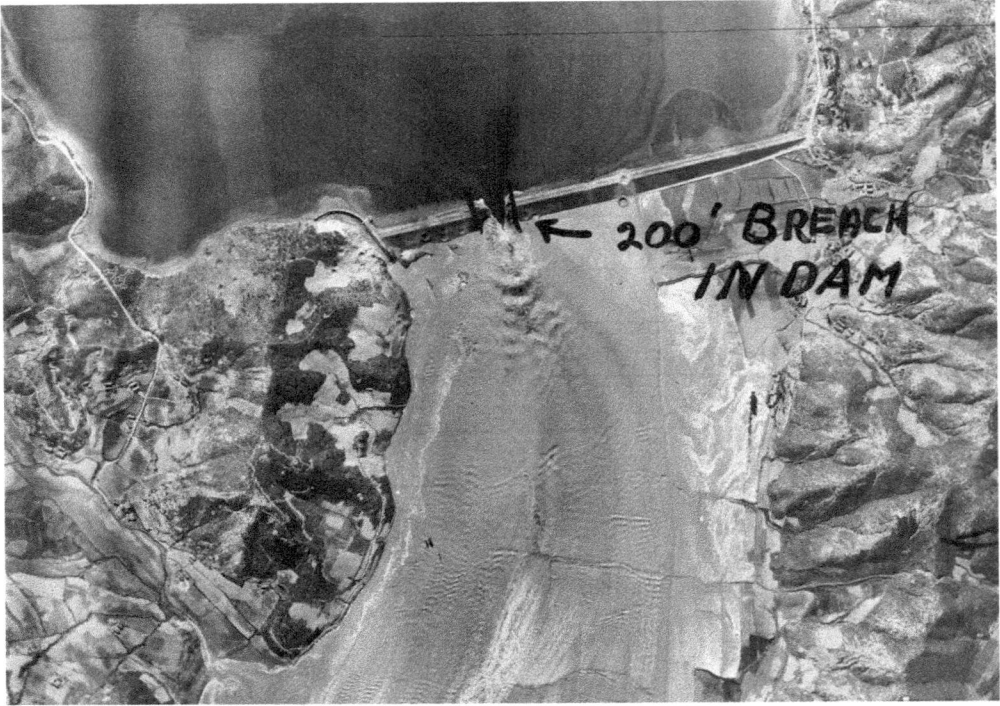

Top on 16 May 1953, F-84E's attacked the Chosan irrigation dam. The first wave of 24 aircraft attained underwater hits, causing a 'tidal wave' of water to pour over the spillway. A second wave of F-84E's then attacked, achieving "direct hit" in a concentrated area, causing a breach in the eastern wall of the dam. Above: Under the code name Operation 'High Tide', aerial refueled missions involving F-84's flying non-stop from Japan to bomb targets in North Korea, commenced in May 1952, when 12 F-84E's flew from Japan and bombed targets in North Korea, refueled by KB-29 tanker aircraft. USAF

This F-84 is displayed at National Museum of the United States Air Force, Dayton, Ohio. USAF

The F-84G began to equip units of the Fifth Air Force in the Far East during summer 1952, with increasing numbers becoming available by September that year. The 9th FBS (Fighter Bomber Squadron) of the 49th Wing was transferred from Korea to Japan in December 1952 to allow the crews to be trained in the delivery of tactical atomic bombs.

The in-flight refueling capability of the F-84G was used operationally during the later stages of the Korean War. In March 1953, F-84G's attacked the industrial centre at Chonjin located on North Korea's east coast, some 40 miles south of the Manchurian North Korean border.

F-84's continued to fly attack missions until the ceasefire was signed bringing the conflict to an end on 27 July 1953. During the Korean War Fifth Air Force lost 335 F-84D/E/G's, the majority of which were brought down by ground fire ranging from small arms to anti-aircraft guns, although numbers were destroyed in air combat operations and accidents. The US Department of Defense lists the numbers lost as 260, the disparity probably being in regards to aircraft lost through operational accidents and non-operational causes. In addition, large numbers of aircraft received varying degrees of combat damage ranging from light to heavy resulting in some aircraft being written off.

APPENDICES

Appendix I

Main variants
XP-84: 3 prototype aircraft procured by the USAAF
YP-84A: 15 aircraft procured for service evaluation
P-84A: cancelled before any production
F-84B: 226 of this variant were delivered as the first production variant
F-84C: 191 aircraft delivered featuring some improvements over the F-84B
F-84D: 154 delivered featuring new wings and a longer fuselage.
F-84E: 843 F-84E's were delivered
F-84G: The most numerous variant with 3,025 aircraft produced. This variant was an improved F-84D and 1,936 of the 3,025 production examples were delivered to NATO nations under the MSP (Mutual Security Program)

Appendix II

F-84B

Engines: 1 Allison J35-A-15 axial turbojet rated at 3,750 lb static thrust at sea level
Length: 37-ft 2-in
Height: 12-ft 8-in
Wingspan: 36-ft 4-in
Weights: 13,465-lb (combat) and 19,689-lb max take-off
Maximum speed: 521 nautical miles per hour
Armament: 6 x 0.50 caliber machine guns; four in upper nose and 2 in wings. Maximum bomb load, 2000-lb

F-84E

Engines: 1 Allison J35-A-17 rated at 4,900-lb thrust
Length: 38-ft 6-in
Height: 12-ft 7-in
Wingspan: 36-ft 5-in
Weights: 15,227-lb fully loaded
Maximum speed: 521 nautical miles per hour
Cruising speed: 485-mph
Service ceiling: 43,240-ft
Range: 1,485-miles
Armament: 6 x 0.50 caliber machine guns; four in upper nose and 2 in wings. Maximum bomb load, 2000-lb

Wing area . . . . 260 sq ft   Wing Section R4, 45-1512-.9
Aspect Ratio . . . . 5.1   M.A.C. . . . . 88.74 in.

185 gal   81 gal   124 gal

130 gal   9.5 gal   81 gal   185 gal

Pressurized Area

Fuel   Oil

Wing area . . . . 260 sq ft   Wing Section R4, 45-1512-.9
Aspect Ratio . . . . 5.1   M.A.C. . . . . 88.74 in.

185 gal   81 gal   124 gal

9.5 gal

130 gal   81 gal   185 gal

Pressurized Area

Fuel   Oil

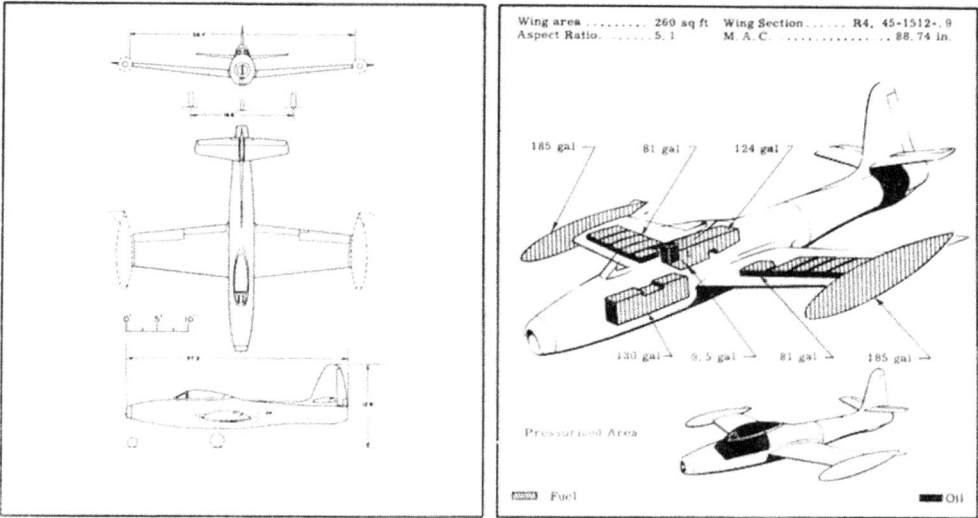

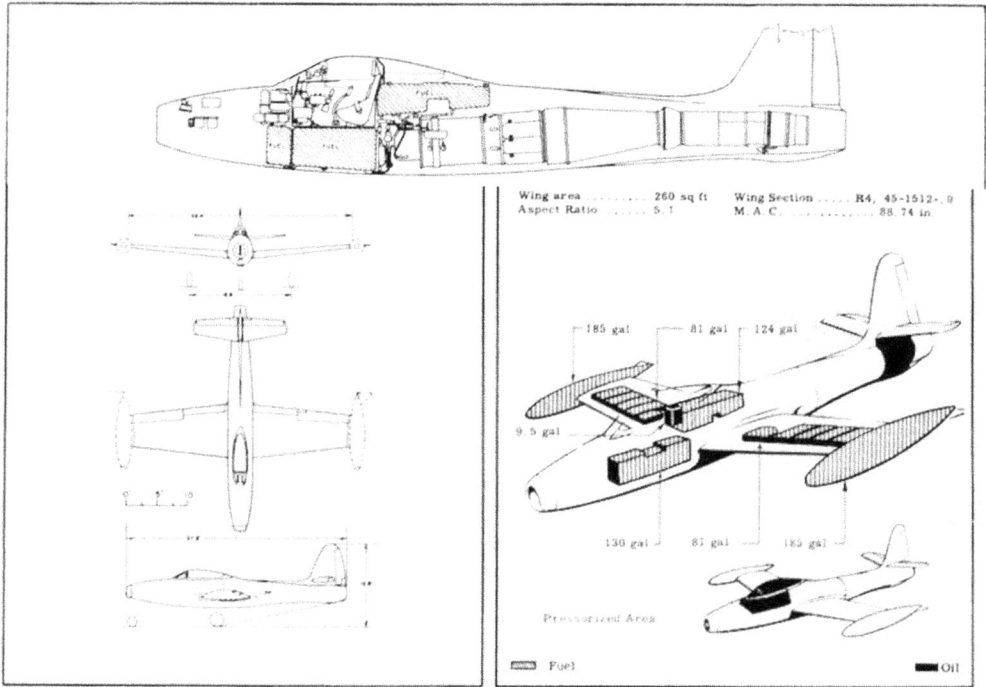

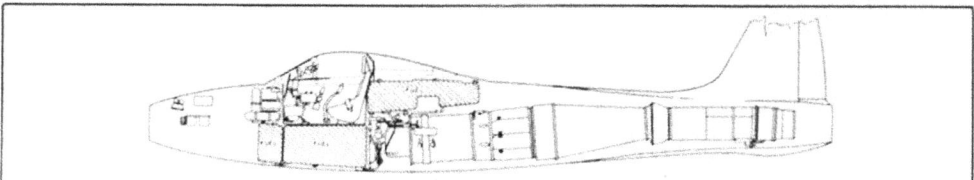

**F-84C May 1950 (top) and F-84D January 1950 (above) USAF Standard Aircraft
Characteristics.** USAF

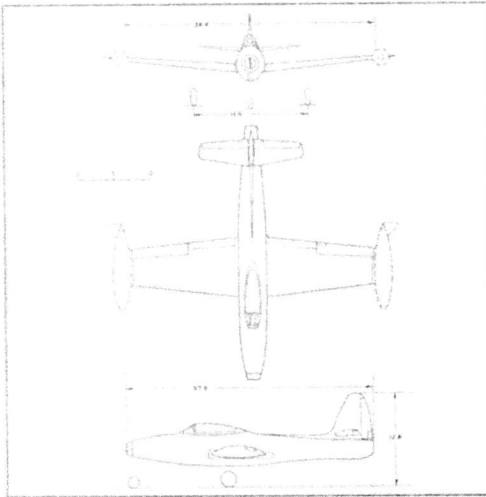

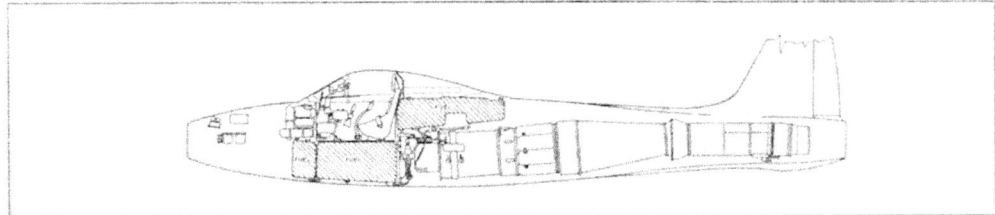

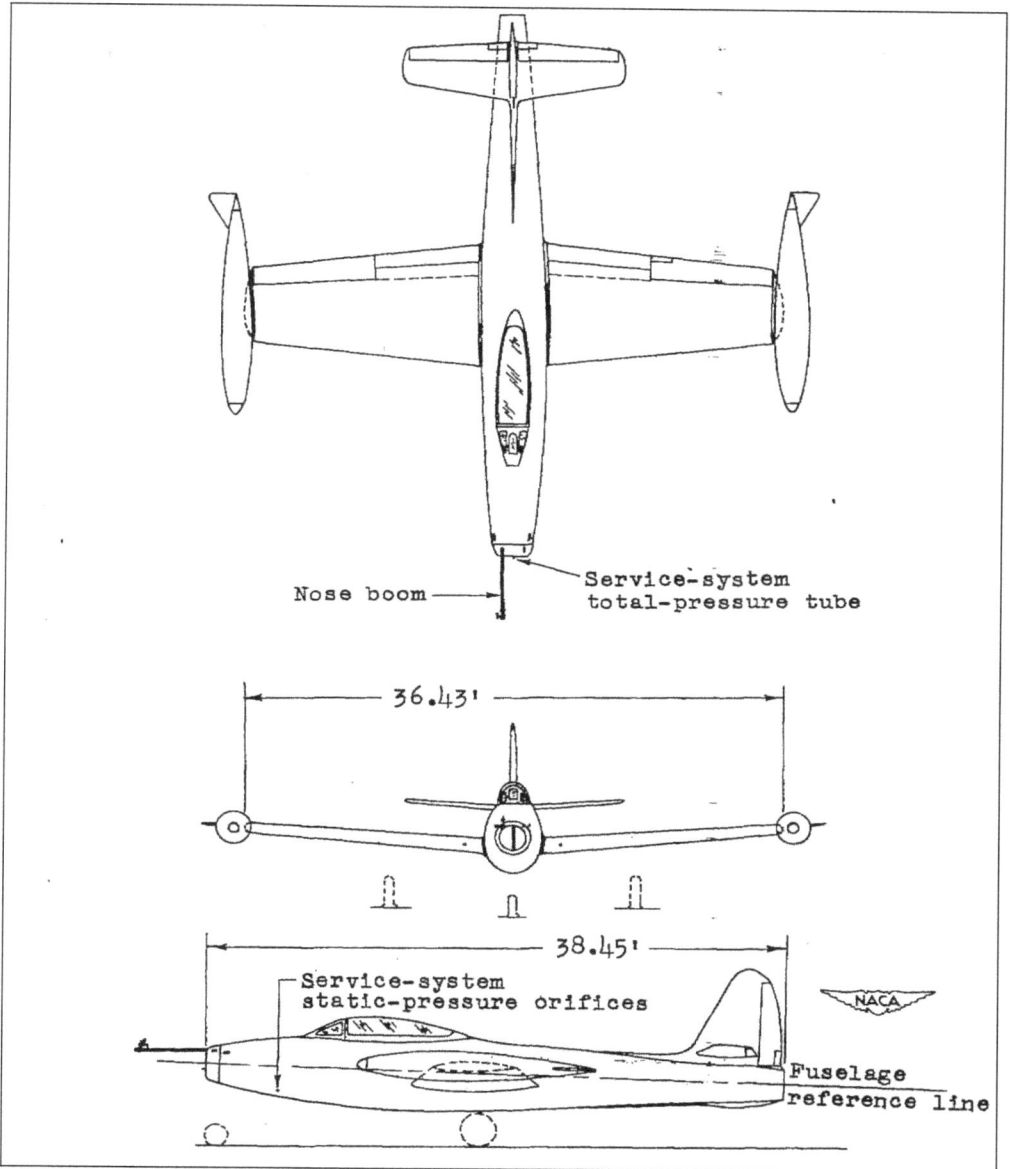

Nose boom

Service-system
total-pressure tube

36.43'

38.45'

Service-system
static-pressure orifices

NACA

Fuselage
reference line

Previous page top: F-84D (Modified) 2 November 1951. Previous page bottom: F-84E Block 25 & 30, 18 July 1951 USAF Standard Aircraft Characteristics. USAF

This page: Three-view general arrangement drawing of an F-84G aircraft with test instrumentation fitted for Squadron Operational Training. NACA/NASA

GLOSSARY

AAA	Anti Aircraft Artillery
AFB	Air Force Base
AFFTC	Air Force Flight Test Centre
ANG	Air National Guard
B	Bomber
CCAF	Chinese Communist Air Force
DoD	Department of Defence
F	Fighter
FB	Fighter Bomber
FBW	Fighter Bomber Wing
FEG	Fighter Escort Group
FG	Fighter Group
FICON	Fighter Conveyor
FIG	Fighter Interception Group
FIS	Fighter Interception Squadron
GOR	General Operational Requirement
HSFS	High Speed Flight Station
HVAR	High Velocity Aircraft Rockets
MAP	Military Assistance Program
MDAP	Mutual Defence Assistance Program
MPH	Miles per Hour
MSP	Mutual Security Program
NACA	National Advisory Committee on Aeronautics
NASA	National Aeronautic and Space Administration
NATO	North Atlantic Treaty Organisation
P	Pursuit
SAC	Strategic Air Command
S/N	Serial Number
TAC	Tactical Air Command
US	United States
USAAF	United States Army Air Force
USAF	United States Air Force
USN	United States Navy
XF	Experimental Fighter
XP	Experimental Pursuit

Copyright © 2013 Hugh Harkins

All rights reserved.

ISBN: 1-903630-61-4
ISBN-13: 978-1-903630-61-7

www.ingramcontent.com/pod-product-compliance
Lightning Source LLC
Chambersburg PA
CBHW080532030426
42337CB00023B/4699

* 9 7 8 1 9 0 3 6 3 0 6 1 7 *